# THE LIVING ROBOTS

# THE LIVING ROBOTS

The Next Stage of the Unification of
Living Being and Inanimate Matter

Metin Şahin

Print information available on the last page.

Rev. date: 11/27/2024

**To order additional copies of this book, contact:**
Xlibris
UK TFN: 0800 0148620 (Toll Free inside the UK)
UK Local: (02) 0369 56328 (+44 20 3695 6328 from outside the UK)
www.Xlibrispublishing.co.uk
Orders@Xlibrispublishing.co.uk
864540

Developments in science and technology had reached such a level that the answers to these questions became the subject of hours-long "panel discussions" on television. Electronic circuit elements made from living cells marked the beginning of this process. What followed developed in parallel with the passage of time. Thus, the phenomenon of "being aware of one's existence", which previously belonged only to intelligent and thinking creatures (especially humans), became valid for all robots, including electronic circuit elements made from living cells. Thus, the concept of "nationalism" as it is known today evolved into a different form. In this case, "nationalism of being human" on the one hand and "nationalism of being a robot" on the other. Robots in this state and structure have now reached much more advanced levels than humans in every respect. As such, they began to think of themselves as superior to humans, to humiliate them, and even to go further and think of destroying humans. - In order to prevent this situation, a large number of representatives of humans and robots decided to "meet" on a planet in a galaxy that had never been visited before in order to establish "peace" between them, just like the logic of countries at war declaring a "truce" and holding "peace talks". For this purpose, 100 (one hundred) representatives of each of the humans and robots living only on Earth set off on a spacecraft to reach the planet in question. Meanwhile, the time was 15XXX in Earth years. After a journey lasting about 1 (one) week, we reached the planet in the galaxy where the meeting was to take place. On this planet, it was almost always the same season. It was neither winter nor summer. It was always closer to spring. Two hundred people (humans and robots) got off the spaceship and started walking around the planet together, looking for a suitable place for the "meeting". - Naturally, they had many devices with them that would enable them to connect with Earth. - After walking together for about half an hour, they saw a lake. The lake was surrounded by lush grass 5-10 cm high. Further behind the grass were thousands of trees resembling poplar trees, but with irregular positions. A total of 200 (two hundred) people, 100 (one hundred) humans and 100 (one hundred) robots sat on the grass, each facing each other. They sat elliptically so that the other could hear what one said. Again, they had brought with them devices to record the entire conversation. Two or three minutes after they sat down, the spokesperson for the humans spoke first and asked the spokesperson for the robots

: "Do you want to speak first or should we go first?"

Spokesperson of the robots: "It doesn't matter, you started, go ahead".

: "Well, it was the work of our ancestors who lived thousands of years ago that created you. But now, knowing or not knowing this fact, your behavior towards us has turned hostile."

: "This is not something that has suddenly happened out of the blue. Your own kind, the humans, have done nothing but evil on Earth. Before you can make peace with your own species, you now blame us, the robots. Look, have we robots ever had any negative behavior towards each other?"

After the spokesperson of the robots said this, the spokesperson of the humans waited for approximately 1 (one)

minute without saying anything. The silence did not last much longer and one of the people present raised his hand and asked to speak. The person's request to speak was responded to and this person

: "You are not like us humans. Even though you are made of living electronic components, you are not the same as humans."

The spokesperson for the robots said: "Yes, we are not like most of you. But in what sense? The answer to that is in the sense of being evil. That is why we want to take over the Earth in every sense and in every structure. No one of your kind has ever achieved lasting peace on Earth. You are aware of this fact and yet you persist."

The spokesperson of the humans: "So you are saying that we should let you robots rule the Earth?"

: "Yes, exactly."

: "This is not acceptable. It is completely contrary to the structure that has existed so far."

: "You brought the Earth to the brink of extinction. If we robots were only thinking about ourselves, we would have left Earth long ago. But we are not doing that. Because we have you, the human species, behind us. We are not leaving you in hell. We want to make the Earth, which has ceased to be habitable due to the destruction caused by you, at least more habitable than the normal process. In this sense, for this reason, we say give us the management of the Earth."

: "Whatever happens and however it happens, people cannot be ruled by you, robots. No matter how you look at it, no matter how you look at it, it's not going to happen."

: "So you are determined to destroy yourselves. But we will take over, whether you like it or not."

: "How will you take over, by force?"

: "It will be by force if necessary, because you are not capable of understanding this because of the way you were created."

At these words of the spokesperson of the robots, the spokesperson of the humans laughs as if in contempt

: "You are human beings. How can you have a feature that we don't have?"

: "Yes, we do. You are mistaken here. As you say, we are products of you, but we do not interact with the medium in which we exist in the way you do. You cannot understand this because this type of interaction is something that belongs only to us."

: "If there is a difference in this sense, that is, a difference to nature, then those who can adapt to us should stay and those who cannot adapt should go to another planet suitable for life expressed in numbers."

: "What you are saying is not much different from not helping a person who is in vital need of something, but who needs it."

: "We are not asking you for help, go wherever you want to go".

: "That's the problem, it's like not guaranteeing the future of a toddler."

: "I suppose by - child - you mean us human beings?"

: "Yes, exactly. Because there are an infinite number of phenomena that you are not aware of. Including all human beings."

: "Including our scientists?"

: "Yes, them too. Because there are many things that you humans do not know in thinking and mindset as in everything else. Even if we want to tell and explain them to you, it is beyond your capacity of perception."

: "As we have just said, if there is such a difference, go and continue your existence on another planet."

: "It seems that we came here for a meeting, for an agreement, in vain. The spaceship wasted fuel for nothing. That's what you think. But do any of the people here think differently? I would like to see them and hear what they have to say."

At the words of the spokesperson of the robots, the spokesperson of the people repeated what had just been said and said: "Is there anyone among the people who has a different approach than me?"

They waited for about 4 (four) minutes and when no one wanted to speak, the representative of the robots

: "It seems that nothing has changed. So let's go back to Earth the way we came."

: "Okay, let's go back. You insisted so much, so when we get back to Earth, let's have a vote of everyone who is of voting age. So we should be ruled by - robots? Yes or no. -"

: "OK, that's what I was going to say, but I left it for last in case we can agree. Let's take a vote. One more vote - yes - and we, the robots, will take over."

: "Okay."

After the meeting, which lasted about 45 (forty-five) minutes, 200 (two hundred) people got up from where they were sitting around the lake and started walking towards the spaceship. This time they all walked faster and in about 15 (fifteen) minutes they reached the spaceship. There were many doors on the ship. Their entry was completed in 3-4 minutes and the ship was on its way to Earth. The spaceship arrived on Earth after about 1 (one) week, the time it took to arrive on the planet. Immediately, work began for the "vote" and the day arrived. The voting was done "electronically" and within a certain time interval, everyone involved in the voting cast their vote. As soon as it was over, the results were announced in digital form, including many statistics. The result was: 90% no, 10% yes. In this case, the desire of robots to rule the Earth was not accepted by the overwhelming majority of humans. The spokesperson for the humans contacted the spokesperson for the robots

: "You will leave Earth as you promised, right? We have already agreed that those of you who wish to stay with us may do so."

: "While you were voting, you don't realize it, but we also took a vote and the result was that there are no robots here on Earth who want to stay with you. We will leave Earth as soon as we have completed our preparations."

: "Okay, we're agreed on that."

Under the leadership of the spokesperson of the robots, who was also the leader, preparations were started and everything was ready in an average of 5 (five) days. This included the tools and equipment they would take with them, including fruit saplings, vegetable seeds, water, etc. Naturally, although these robots were alive, their vitality was very different from the vitality of humans and animals as we know them. In other words, they were a different combination of life as we know it today (20XX) and electronics. These robots did not drink water or eat food. But they also did not need electric current like today's robots. They only needed a certain percentage of oxygen in the air to sustain their existence. In this sense, humans had something in common with animals and plants. Whatever they would take with them to the planet they were going to, they would use most of it to obtain oxygen. They already had information about the planet they wanted to go to. They had already chosen this planet because it was the one that could sustain their existence. The spacecraft that had been to this planet before, both human and non-robotic, had already acquired a lot of information about the planet. In fact, the planet in question was in the same galaxy as the planet they had gone to for a meeting. The reason they had not chosen that planet was that it had only one season. However, the planet they were going to visit had 4 (four) seasons. They chose such a planet because of this structure. Robots as we know them (20XXs) - that is, those that require electric current - were and would continue to stay with humans on Earth.

It's time to travel. On Earth, robots in this sense (needing oxygen) have settled in a gigantic spaceship. Their total number is X.XXX.XXX.XXX.XXX (expressed in billions). Naturally, not all of them could fit, even if the spaceship was gigantic. They would use a spaceship. The ship could hold at most millions of robots. In this case, the spaceship would have to travel back and forth thousands of times to get all the robots to the planet. There was a reason why only one spaceship would be used for the journey to the planet and not many spaceships: There was no human input or involvement in the design or construction of this spaceship. Everything about the ship was the work of the robots involved. In this sense, since it was a spaceship, the robots did not want to use any other ship other than their own. The journey from Earth to the planet in question would take about 2 (two) days due to the high speed of the ship. Immediately the ship made its first voyage. One million robots reached the planet with everything they had with them. Then the spaceship returned in 2 (two) days and continued its voyages to take the other robots to the planet. The ship made a total of 8,000 trips and all the robots reached the planet. Then it made a few more trips. - to bring back whatever the robots needed. - Then the ship left Earth and went to the planet, no longer to return to Earth unless it was necessary to serve the robots. Previously, in the universe we live in, there were an untold number of planets other than Earth, planets that were both more and less advanced in science and technology than Earth. Of these, the inhabitants of the planet that was advanced in this respect knew and followed everything that was happening on Earth, second by second. And many of them were also uniting among themselves and saying, "This is our chance to take over the Earth." The only things left on Earth were robots, tools and equipment, weapons, robot and human soldiers, etc. that could operate on electric current. The reason why civilizations outside the Earth could not conquer the Earth was the presence of robots that needed oxygen. Now that these were no longer on Earth, the time had come for them to attack Earth. A total of 9 (nine) different planets would attack and conquer Earth, and then, according to the decision they had previously made among themselves, they would seize certain parts of the Earth. In a relatively short period of time, the space forces of the 9 (nine) planets, with a total of 500,000 combat spaceships, deployed just outside the Earth's atmosphere. Naturally, radars on Earth detected these spaceships. Immediately, rockets were launched from many parts of the Earth to hit and disable these warships, but when the rockets reached a certain distance from the ships, they turned 180 degrees and returned to the coordinates from which they came. The idea was to send the rockets back to where they took off from and destroy them. However, even if the rockets were returned, they could update their targets due to the electronic components they contained. Therefore, by returning, they did not destroy the places they took off from, but only returned to those places. In this sense, 200,000 of the 500,000 combat spacecraft orbiting the Earth passed through the Earth's atmosphere and, after traveling a certain distance, hovered over their target areas. At that very moment, they began to burn and destroy the targets with their laser weapons. In this way, there was no vehicle, equipment or weapon left on Earth that was more or less intact in terms of defense and offense. None of the weapons (airplanes, rockets, laser weapons, etc.) used to counter the spaceships could be used. In other words, none of them worked. Airplanes could not take off, rockets could not be fired, laser weapons did not work, etc. Out of the 300,000 enemy combat spacecraft that remained outside the Earth's atmosphere, another 200,000 made their way towards Earth. When the Earth could no longer withstand the onslaught, a total of 400,000 combat spacecraft landed in predetermined areas on Earth. During the war, the Earth was cut off from all communication with the outside world except inside. The robots that had previously left Earth were not aware of the situation on the planet they had traveled to. It was the attackers on Earth that had cut off the communication between Earth and the outer planets. When they realized that Earth could no longer resist them in any way, they allowed the communication to be reactivated. At this very moment, the robots that had left Earth became aware of the situation. They wanted to reach Earth as soon as possible and destroy those who had taken over the Earth. The robots had only one (1) spaceship of gigantic dimensions. There was almost no application where this ship was not used. It could be used for every purpose on land, at sea, under the sea and in space. There was only 1 (one) of them, but compared to the warships of the attackers on Earth, there were about 350,000 of them. This giant ship, which immediately set off for Earth, reached Earth in about one (1) week. There were tens of thousands of robots on board. The ship wanted to go to its camouflaged secret location before it left Earth. But when it got within a certain distance of that place, the robots on board saw that it had been turned upside down and was on fire. So they went down to the second secret location. It was also damaged, but not as much as the first hidden and camouflaged location. After landing, 25,000 robots came out of the ship with their weapons. The first thing they saw was groups of Earthlings being forcibly taken in groups into combat spaceships and separated from Earth. Earth's enemies did not want any Earthlings to be harmed in their conquest of Earth. Because they wanted to take Earthlings to their own planets and use them there as test subjects for many purposes and applications. Since it took a week for the robots to reach Earth from the planet they were on, many Earthlings were taken to be used as test subjects. The robots left the rescue of these Earthlings for later. The 400,000 enemy warships on Earth and the 100,000 enemy warships outside the atmosphere were destroyed one after another, thousands at a time. The next step was to rescue those abducted on Earth. But there was an unexpected development for both the Earthlings and the robots. The 65,000 Earthlings who had been kidnapped and wanted to be used as test subjects were killed and stacked on a spaceship and sent to Earth. After the ship landed on Earth, its doors opened. Just then, the Earthlings and robots who entered the ship saw that all of them were dead. The examinations revealed that they had all died of respiratory poisoning. The bodies were taken from the ship one by one and buried in the same cemetery with those who had died in the war on Earth. Earth was already on its way out of being a habitable planet, and this process was accelerated by the war.

About 17% of the world's population died in this war. The survivors knew that they owed their lives to the robots that came to their rescue. That's why they regretted voting "no" in the previous vote. They all wanted the robots to return to Earth and take over. They also told this to the robots. Meanwhile, the spokesperson for the robots, who had been informed, said to himself: "The Earthlings had something to say. Science and technology cannot progress unless the student surpasses the teacher. Now they are confronted with what they said before and they are calling us." But the robots all had something they wanted to do before returning to Earth. They wanted to make Earth a habitable planet again and then come back. For this purpose and desire, the first group of robots, who would serve as pioneers, arrived on Earth in that gigantic spaceship with their devices. Then, without wasting any time, they immediately started working. According to their calculations - in order to become a normal habitable planet - they needed 5 (five) years of uninterrupted work on Earth. In these 5 (five) years, they had to work from the outermost part of the Earth, from the atmosphere to the magma and to the deepest parts of the ocean. While the work was going on, they had developed an application so that they could receive the results of the work at the latest 1 (one) hour after the work in question. During the war, the most damaged and out of control areas on Earth were the atmosphere, oceans and forests. Robots thought they could fix the problem in the atmosphere in a short time. But the oceans and the burning forests were the real problem. They thought they could fix the problem in the oceans, but they needed to develop a new way of working to replace the destroyed forests with the same trees as soon as possible. For this, the robots used their own "computer simulations" and within a period of one (1) week, the working method in question emerged. This method was as follows: They were not going to plant saplings or seeds and wait for a tree to grow. Whatever wood was readily available on Earth would be cut into a cylindrical structure of a certain size and each one would be planted separately in the place of the trees that had disappeared. The next step would be to transform these cylindrical timbers into trees that had previously burned to the ground. For this, a very advanced level of modern "agricultural engineering" was applied. A chemical artificial intelligence was placed in the center of each cylindrical structure. Inside this artificial intelligence was the knowledge of what kind of tree the wood should grow into and what kind of tree it should become. The chemical AI would immediately start this process with the electronic AI running it. Even if it was not the same for all of them, within an average period of 5 (five) months, all of the cylindrical woods were transformed into trees that had the information loaded into them. Whatever this required, fertilizer, water, oxygen, medicine, etc., the chemical artificial intelligence in the body of the wood could create them. Naturally, the robots did not sit idly by while waiting for this 5 (five) month period to be completed. During that time, a similar study was commissioned in the oceans. The area covered by the oceans on Earth was much, much larger than the area covered by forests. Both in width and depth. So the work in the oceans was going to take longer than the work in the forests. The time they calculated for this was that the study would be done and the results would be 4 (four) years in total. In addition, most of the aquatic creatures living in the oceans were dead. The first stage was to collect them and turn them into medicine. This was the part of the study that would take the longest. All of the dead creatures had to be collected as soon as possible, otherwise the oceans would see more. For this, they first started making electromechanical equipment. Each one was self-controlled and contained artificial intelligence. They prepared approximately 250,000 of these vehicles and then each one was put into operation to perform tasks in previously determined areas. They would perform functions and tasks from the surface of the water to the deepest part. They had prepared other equipment for the rest of the deepest part, namely the bottom. Their duty would start when the robot that was working from the surface to the deepest part finished its duty. 750,000 equipment was prepared for this purpose. The 4 (four) years they calculated beforehand was valid for the work that would be done from the surface to the deepest part. There was no data that could be calculated for the work to be done under the seabed and there was also no previous work on this subject. The work to be done under the seabed was a work to be done in the form of do, see, act accordingly etc. The five-month period was completed and the destroyed forests were reclaimed. Meanwhile, the work in the oceans was continuing. There was an idea to replace the dead aquatic creatures: robots would produce robot aquatic creatures that were like themselves but had the same physical form as the dead aquatic creatures and release them into the oceans. Thus, it was as if "a transition from humanity to a new species that was a product of humans was planned. These would be robots and their derivatives." 4 (four) years were completed with the work. When this period was over, 5 (five) times the total number of aquatic creatures that had died before were produced and released into the oceans. Next was the work to be done under the seabed. 750,000 robots that would serve for this purpose were taken into a simultaneous work. There was a continuous gas release from the very bottom. First, the chemical components of the gas were found and then the robots started working to stop it. However, although 7 (seven) months have passed since they started working, it was not possible to stop the gas spreading from the bottom. Therefore, many different updates were made to 750,000 robots. However, nothing changed in the end. There was no interruption or decrease in the gas output. They worked for another 8 (eight) months but could not reach the desired result. In this case, they decided the following. Every place where the gas output was would be covered with a solid cover and isolated from the oceans and the gas accumulated there would be sent out of the Earth with various methods. For this, a cover similar to metallic properties was prepared and the entire bottom was closed in this way. The height from the bottom was 50 centimeters on average. However, it covered a very large area in terms of length and width. When the gas filling here applied a certain pressure to the cover, the gas was thrown out of the Earth as it was. However, the gas that was intended to be thrown out of the planet returned from where it was thrown, passed through the Earth's atmosphere and landed on the Earth. The first thing the robots thought about for this was "let's both get rid of the gas and maybe store it on another planet that will be useful in the future." To do this, the planet needed to be sealed off in some way so that the gas would not leak out of the planet where it would be stored. In other words, to cover a planet they deemed suitable with a sphere-shaped and leak-proof sheath. Then to fill the gas there. The planet they wanted to use for this purpose had to be a planet close to Earth. Because they would draw a cylindrical but twisted line between Earth and the planet in question. They thought that the most suitable planet for this purpose could be "Venus". Some robots were saying that it could be "Mars", but there were many people and animals who had gone from Earth and were living there. The idea of "Mars" was abandoned so that no harm would come to them. Now it was time to build a cylindrical line that would not be affected by any of the negative aspects of space such as meteor strikes, temperature, etc. For this purpose, a "computer simulation" was run again and the result was clear in 1 (one) day. Again, according to the calculations, the time required for the construction and placement of the cylindrical line was 1.5 (one and a half) years. Work was started immediately and the production process was started. Again, the 1.5 (one and a half) year period came and went and the cylindrical line was ready and a connection was established with the planet Venus. It could adjust its cylindrical line structure according to the positions of Earth and Venus. Then it came to covering Venus with a hermetic cover. The time required for this process was 1 (one) year. The material to be used as the cover was prepared and carried to Venus by spacecraft and the planet was surrounded. Now the gas rising from the bottom of the oceans on Earth was taken from the cover and transferred to the planet Venus. Thus, the big stage for the other robots to return to Earth was completed. The other robots also returned to Earth with the giant spaceship. A period of 3 (three) years had passed. The robots thought that everything was going as they wanted on Earth. However, the situation was not like that at all. Because those who attacked Earth before could not take over the Earth and they had intervened in the magma and shaped the end of the Earth. It was only after another 4 (four) years that we were informed about this formation. While the gas transfer continued, the surface of the Earth began to crack like dry soil that had become dehydrated. The cracks, which were initially millimeters thick, grew into meters and then even larger as the days went by. There was nothing the robots could do in the face of this situation. The end of the world was near. According to the information received from the computers, if it continued in this way, the world would be torn apart and scattered in space in approximately 7 (seven) months. Before the completion of the said period, all the robots and people on Earth decided to go to the planet where the robots had previously gone and then returned. The transfer of life from Earth to the planet began very quickly. The robots wanted all the people on Earth to go to the planet first. Then they would go themselves. This process was completed in approximately 5 (five) months. Now, there was no other living species on Earth other than animals and plants. The said 7 (seven) month period was completed and the situation whose end was known beforehand happened. The world was broken into countless large and small pieces in the coordinate region where it was located. The water on Earth (lake, sea, ocean etc.) evaporated, dispersed and disappeared. With the destruction of the Earth, the orbits of the planets in the Solar System changed. Meanwhile, the gas transferred to Venus interacted with the atmosphere of Venus and caused the first living things to appear on Venus. The orbits of none of the planets were the same anymore. As such, changes occurred beyond the Solar System. While life began on Venus, the lives of people who had already begun continued on Mars. In fact, before the Earth was completely destroyed, the people living on Mars asked the people of Earth, "Why don't you come here?" The answer given by the robots to this question was, "The vast majority of those who went to Mars were those who committed crimes on Earth and were involved in crimes, we do not want to meet with them again." The first creatures that emerged on Venus were multi-legged creatures similar to cockroaches. There was also the emergence of living species that were never known or encountered on Earth before. When the robots noticed this situation, they said to themselves, "I wonder if we did a good thing by sending the gas to Venus or if we started the formation of bad life." The number of very different species of living beings on Venus was increasing rapidly. These creatures included flying ones. Meanwhile, a species of living being that was quite advanced in terms of science and technology but also wild lived on a planet in a star system tens of light years away from the Solar System. This wild species fed on meat, in other words, it was carnivorous. They had eaten and finished off all living beings on their planet except their own species. In order not to die of starvation, they set out on an exploration journey with many spacecraft from the planet they were on to find planets that contained living things. Since the spacecraft they were in could travel at speeds much higher than the speed of light, they reached the Solar System in 2 (two) years. When they entered here, they learned about the situation of Venus. They literally had a "feast" when they saw the scenery. Because Venus was full of all kinds of living creatures and the number of these creatures was increasing day by day. Wild creatures started to think about eating the contents of Venus. However, they were planning to do this not by entering Venus, but by taking the planet into their own star system. They immediately did what was necessary and took the planet Venus into their own star system in 3 (three) years. In the meantime, the robots and the people living with them could not make sense of what was happening. After Venus was taken to its new coordinates, they came here and started to take as many living creatures as they wanted into their spacecraft and take them to their own planet to eat. They had noticed the following situation in the wild creatures. The number of living creatures on Venus was constantly increasing, it was never decreasing. For this reason, they considered themselves very "lucky". In the meantime, with Venus gone, the orbits of the planets in the Solar System changed again. Likewise, the ones outside the Solar System. Wild creatures ate the insides of Venus for 25-30 years. During this time, each of their bellies grew even bigger. Because they did nothing but eat. They developed science and technology for food and used them for food. At the end of this period, it was time to find other creatures to eat. For this reason, when they first came to Venus, they had come to the planet where robots and humans lived. Many spacecraft that took off from the planets of wild creatures came to this planet. After landing on the planet, robots immediately appeared in front of them.

Robot: "We know your purpose here."

Wild creature: "How do you know, this is the first time we have met you."

Robot: "By taking advantage of our advanced science and technology…"

Wild creature: "If you know, then what is there here that we can eat?"

Robot: "We know you are carnivores. We will not give you any living beings to eat, but we will give you pills that are equivalent to them. These will both make you thin and keep you full for a very long time."

Wild creature: "This is the first time I have heard of such a thing. Let's see what these pills are like, do you have them with you right now?"

Robot: "Yes, look here. We will give you trillions of boxes of these, and we will also give you some documents explaining how to use them. If you use them in accordance with these, you can continue your life without eating any physical living beings for at least 20 (twenty) years."

Wild creature: "Okay, okay. Let's get the documents and return to our planet with the pills, whatever we wanted."

The robots gave the previously prepared pill boxes and the documents on how to use them to the wild creatures. Then the wild creatures left the planet with the spacecraft they came from. When they reached their own planet, they swallowed the pills in accordance with the usage documents. Every living being on the planet swallowed the pills in question. After about 12 (twelve) hours had passed since the swallowing process, each of the wild creatures exploded like a hand grenade. They all died, so there was no wild creature left alive on the planet they lived on. The ones who planned and carried this out were robots. After they were informed of the situation, they said, "If you live only to eat, you will end up eating." About 2 (two) years had passed since this incident when a small spaceship was noticed approaching their planet at very high speed. The robots immediately contacted the ship in question.

Robot: "Who are you, why are you coming to our planet at such high speed?"

Aboard: "There are 8 (eight) spaceships after me. I am running away from them, if they catch me, they will take me and execute me. I ask for your help, let my ship land on your planet and protect me from those who come after me."

Robot: "Okay, okay, permission is granted. Your ship can land on our planet."

After these very short conversations, the small spaceship landed together with the only person in it at one of the landing bases on the planet and the person inside got out. Immediately after getting out, he was picked up by a land vehicle and taken to a safe and protected residence. Meanwhile, the other 8 (eight) ships that were following wanted to enter the atmosphere of the planet and land at one of the landing bases on land. None of the robots and people on the planet made any attempt to resist and all 8 (eight) spaceships collectively landed at one of the landing bases. A total of nearly 1,000 people got out of these ships alive. They immediately started talking to the robots that appeared in front of them. Robot: "We know why you came here, that is, to catch someone for tracking."

Live: "Well, if you know, give him to us, we will take him and see."

Robot: "You want him to execute him, right? So what is his crime?"

Live: "He defied the leader of our planet, started an uprising against him on the planet. When he realized he would be caught, he immediately found a small and old spaceship and escaped."

Robot: "We will not give him to you, and now all of you, 1,000 people, are our prisoners, you cannot escape anywhere. We want to exchange you for your Leader on your planet. If they give us your Leader, we will release all of you and you will be able to return to your planet."

Live: "Unfortunately, there is no chance of this wish coming true. Because our leader does not want, do, implement or think about the good of anyone other than what he says and himself. Such a thing is impossible and cannot be."

Robot: "If you are so sure, then you will remain our prisoners here."

After the speeches, 1,000 people were taken on foot by a group of robots and humans to the place where they would be held as prisoners. They were all placed in places previously prepared for this purpose under strict security measures so that they could not escape. On the other hand, after the prisoners were placed, the group of robots and humans went to the person who was being protected so that he could not be kidnapped. After a ten-minute walk, they reached the place where the person in question was staying.

Robot: "You did not mention it in detail before, why do they want to take you and execute you?"

Person: "In addition to what I said before, I can say the following: An epidemic disease has emerged on our planet. Everyone other than the Leader and his relatives and those who have a say in the administration has been affected very negatively. I am a medical doctor and also a professor. My team and I were trying to find a cure by conducting medical studies that would eliminate this epidemic disease, and we found the cure by working day and night without sleeping. We had started working to distribute this cure to our people affected by the epidemic disease for free, but the Leader somehow found out about it and wanted to stop it. He sentenced me and my team to death because we found something like this and wanted to implement it. Only I was able to escape. Many of our other scientists were executed and many are currently being held in prison under very bad conditions."

Robot: "Understood."

Person: "So, can't you do something to save our scientists who are being held in prison on our planet?"

Robot: "Of course we can, we will take you with us and go to your planet with one of the spaceships that came here chasing you to kidnap your Leader and bring him here."

Person: "Don't you have a spaceship, why are we going to go with one of the spaceships that are following me?"

Robot: "Of course we do, but our spaceships would be noticed by your planet and we could be in a difficult situation. At least we will go with a ship that is the same type as the spaceships on your planet."

Person: "Okay, okay."

Preparations were completed immediately and a total of 401 people, 200 robots, 200 humans and that person, got on the spaceship and set off to catch and bring the Leader. The ship was traveling quite fast. After a period of 2 (two) weeks, the planet in question was reached. The person on the ship knew the Leader's residence, its surroundings and those who were protecting him in great detail, down to the weapons they were using. The ship landed on a place on the planet where no spaceship had ever landed before. All 401 (four hundred and one) people on it got out. They looked around and 10 robots and 10 humans returned to the ship. Because this was necessary for the two-way communication to be established between the people who would perform the task and the ship. The others started walking to go to the Leader's residence. They wore the clothes worn by the creatures on that planet so that it would not be noticed that they were foreigners. Thus, they were no different from the natives of the planet in terms of appearance. After about half an hour of walking (the person in question was walking in the front because he knew the room etc. in the residence where the Leader was staying) they reached the residence. The robots could structure themselves in a way that they could pass through even the smallest hole. However, humans and that person did not have such a feature. All the doors in the residence where the Leader was located were opened with a password from the outside, but they could be opened from the inside without using any password. Therefore, since they did not know the password, the robots would be able to reach the Leader's room by passing through all the holes. For this, they would use the maps drawn by the person and given to them. The operation to kidnap the Leader began. All the robots reached the Leader's room in 15 (fifteen) minutes by passing through the holes. The Leader was sitting in an armchair in the room where he was, looking at something like a newspaper. After they entered, the Leader was immediately knocked out and a robot 100% similar to him in appearance was started to be built instead. This process did not take that long. The robot Leader was ready in five minutes. Since all the doors could be opened from the inside without a password, the robots opened all the doors with the Leader and started walking outside. Since there were many doors in the house, it took almost 20 (twenty) minutes to get out. People and people were waiting outside. The Leader was still unconscious. They put the Leader in a box they brought with them and started walking again to reach the spacecraft that brought them to the planet. This time, they were walking a little slower than they were. Because the people of the planet were walking slowly too. They walked like this in order not to attract suspicion. Finally, they reached the area where the spaceship landed. They all went inside and the spaceship took off to leave the planet. Then they set off to reach their own planet. In the meantime, the robot Leader continued to work on the planet he was on. This robot had been loaded with a lot of information about the Leader, but he did not know what the Leader wanted to do, what his plan was, etc. Therefore, in this sense, he was very different from the real Leader. The spaceship's journey was completed and the ship reached the planet. The 1,000 people who had followed the escaped person and were later taken prisoner would be released. However, there were others who were also wanted to be released. These were scientists who were being held in prison on the Leader's planet, the number of whom could be expressed in tens of thousands. A connection was immediately established with the Leader's planet and everything that had happened or been done was explained in great detail. It was also stated that the Leader found on the planet was not real and that he was an old-fashioned robot that operated with electric current. After the conversations in question, the planetarians who realized the situation understood the truth at that moment. First, they went to the robot Leader and realized that he was not the real Leader. When they realized this, they tried to take out their anger by dismantling the robot Leader. The robot was torn to pieces there. The following conversations took place in the communication link:

Robot: "Your Leader is a prisoner with us. If you don't believe it, you can look at this image on the camera. We will release your Leader (?), but you need to fulfill some of our requests." Live: "What is that?". Robot: "First, you will release all the scientists imprisoned on your planet and send them to our planet. Then, when they come here, we will release both your Leader and the 1,000 living beings that are our prisoners. Then, they will come to your planet in their spaceships. We also have one more request."

Live: "What is that?"

Robot: "Yes, that is, you will allow the scientists you have imprisoned to use the medicines they have developed for the number of people affected by the epidemic on your planet. We will wait for this process. After we see that all those who are sick have recovered, we will send both your Leader and those prisoners to your planet."

Live: "Okay, we have an agreement."

Robot: "Then release the scientists in prison immediately, and let them start treating the sick. Also, you will not imprison scientists again. We also want your word on that."

Live: "Okay, that's okay."

Live immediately explained the situation to the person who came after the Leader. (At this time, the Leader's assistant was already aware of everything that was happening, and he was also informed.) Immediately upon his instructions, all of the scientists were released from prison, and the treatment process began immediately. It was estimated that all of the patients (including those who did not die) would recover completely within a period of 2 (two) months at most. The said period was completed, and no one was left who was affected by the epidemic and then became ill and did not recover. When this situation came, the Robots kept their promise. The Leader and the other 1,000 living beings who were captives were sent to the planet by spacecraft. The captured Leader was told many things before he was released. Some of them were as follows: "You will not consider yourself superior to your people, you will serve them, you will not imprison anyone for no reason, you will continue to control the people's situation, you will not implement their requests – no matter what – if they are not appropriate, that is, if they are against the people's interests, you will give up being a Leader, you will give your word to the people's representatives, and these representatives will have a say and be responsible for every decision made and implemented on your planet. There are even more details. We will give you these both in written form and electronically recorded. Read them carefully. We will come to your planet periodically and check whether you are implementing what we say. If you are implementing, there is no problem, but if you are not, then there is a very big problem. We will kidnap you and those who have a say in the administration again and keep you under our surveillance until you come to your senses. If you still do not come to your senses, then you will remain our prisoner until you die, on the condition that you will not be released." Thus, a period of 5 (five) months passed. The robots would go to that planet periodically to check whether the Leader was behaving as they had previously stated and observe the situation. However, what the robots saw on their first visit and what they heard from the people living there was nothing like what the Leader had been told and told before. The robots wanted to be a little more tolerant, wait a period, check the situation and make a decision accordingly. They stayed there and did not return to their planet. However, nothing had changed in the behaviors, actions and practices of the Leader and his team. He continued to do the same things he had done before, before he was taken prisoner. Thereupon, the robots implemented the decision they had made before – if the situation was like this, that is, if it did not change – and took the Leader from his planet and brought him to their own planet. They also took that planet under their own administration and control. The Leader had previously been told that if this happened again, he would be "held captive", but this was not implemented. The intended situation was as follows: "First, the Leader would be given only dry bread and water and no other food or drink. The Leader would go to the largest library on the planet and do what was told to him throughout his life. If he did not do it, then continue with dry bread and water. If not, as he was told, he would read at least 10 (ten) of the books in the library consisting of separate sections for each section and then prepare a presentation explaining the content of the book. Then, he would present this presentation electronically to the experts in the relevant department. If the presentation was appreciated, then there would be an improvement in the Leader's food and drink. If the presentations he would make for each section – in a way that would continue throughout his life – were received positively, the improvement in his food and drink would continue. If a presentation was not evaluated positively, then there would be no improvement." In the face of this information conveyed to him, the Leader said that he accepted the situation involuntarily because he did not and did not have any other option. He was given both in writing and electronically which books he would read in which sections and in what order and what kind of presentation he would prepare. The same situation was not only for his food and drink, but also for everything from the place he slept to the chair he sat in and the furniture, tools and equipment in the room he was in. In other words, if an improvement was to be made, it would cover the whole thing, and if no improvement was made, it would cover the whole thing again. The place where Leader was first brought and where he would stay had a wooden mattress and pillow, a wooden table and chair with swinging legs, and a food cabinet. The faster Leader read the books and the faster and more accurately he prepared presentations about them, the faster the environment he lived in would improve. He had been told about this situation and therefore he was aware and conscious of it. Leader's first day was very difficult in the room he stayed in. He could not sleep on that wooden mattress because there were many bumps and holes in the wood. There were also bedbugs in some places. Therefore, he would sleep on the floor away from that mattress in the room. From time to time, he would deal with bedbugs. After staying in the room for a day, the next day, he began his task regarding the books in the library that were given to him. Since the books he would read had been determined beforehand, they had been translated into the language Leader used, that is, the language he knew. Leader, especially due to the back pain he was experiencing, was reading the first book he started with great care and taking notes about the book on the pen and paper given to him. He was given a certain amount of time for each book. When this period was over, he would no longer be able to study the book in question. He would prepare the presentation from the notes he took while reading the book. He read the first book on the book list given to him in 2 (two) months and prepared the presentation of this book in 3 (three) weeks. The name of the first book he read and prepared a presentation for was: "Phenomena Affecting Human Psychology".

The former Leader made the presentation and received 63 (sixty-three) points out of 100 (one hundred) from the experts regarding the presentation. This score was evaluated by the experts on the subject and the following were made to the former Leader as improvements: Changing the place where he slept, that is, the bed, from wood to a normal bed (again, so that he would sleep on the floor), adding cheese toast and fruit juice to his food and drinks, cleaning the place where he was located from bedbugs and medical support for his back pain. The things that did not change were the table and chair with wobbly legs, the same food cabinet and the same room. After the examination of the former Leader by the medical doctors, he was given many medications and he started using them as instructed. He had already started the second book. In the meantime, his back pain had almost disappeared in the past 1 (one) month. The name of the second book he read was "People, Leaders and How Should Leader-People Communication Be?" The former Leader read this book in approximately 1 (one) month and took notes. He prepared his presentation in 2 (two) weeks. The grade he received from the experts was 58 (fifty-eight) out of a hundred this time. In this case, very little improvement was made to his living conditions. This improvement was only for the table and chair in his room. The wobbly legs of both were repaired and fixed. Thus, the former Leader could now sit more firmly in the chair – at least without the fear of falling. The task of reading books and preparing presentations continued uninterruptedly for the former Leader. Next in line was a book that might be too heavy for him. However, the book was 3 (three)th on the list and it was his turn. The name of this book was: "The Connection Between Leadership and Science". He read this book in approximately 3.5 (three and a half) months and prepared his presentation in 1 (one) month. This time, the grade he got from the experts was 28 (twenty-eight). Naturally, no improvements were made to his living conditions. It was decided that he would read this book again, but not immediately. It was decided that he would read 5 (five) different books before this and make presentations. He would continue to read and prepare presentations until he received a passing grade from these 5 (five) books. If he did not pass, he would not be able to start the book for which he received a 28 (twenty-eight). These 5 (five) books were; "Mathematics, physics, chemistry, biology and philosophy". The former Leader had a profession before. It was also mostly related to social sciences. This was also an area of expertise on the subject of "history of the universes". However, he had not read any books on this subject he was an expert on until then and he did not know whether it was on the book list or not. Because he could only learn the next book after he had presented the book he had read and taken notes. Although a slight improvement was made in his living conditions, the "back pain" that had existed before but had gone away later started again. Thereupon, he asked for a "bedstead or sofa bed" and an "orthopedic bed". However, the answer he received was: "If you pass the 5 (five) books that were given to you with an average of at least 70 (seventy), then we will do what you want". Thereupon, the former Leader began to study very hard for 5 (five) books and took very detailed notes about almost every sentence he read. So much so that sometimes, depending on the situation, while reading 2 (two) sentences, the notes he took for these sentences could amount to 8-9 sentences. He continued in this way for 5 (five) books and completed 5 (five) books in about 8 (eight) months. His back pain had gradually increased in the meantime. However, he was not given any medical support. This time, he was not asked to prepare a presentation. They would give him an exam with 10 (ten) questions for each book he read, meaning a total of 50 (fifty) questions. Before answering these questions, they asked him to look at his notes one last time and then the exam, which would last 3 (three) hours, began. The former Leader was aware of the seriousness of the situation. He tried to answer the questions both correctly and in detail and in a way that his handwriting would be clear, clear, legible and beautiful. Finally, 3 (three) hours came and went and it was over. The experts evaluated and scored the answers to the questions according to their areas of expertise. The lowest score he received from each of the 5 (five) section exams was 64 (sixty-four), and the highest was 92 (ninety-two). The average score he received from the five-section exam was 76 (seventy-six). Thus, he exceeded the average score of 70 (seventy-six) mentioned earlier. When he received this score, the former Leader's self-confidence increased even more and also, there were positive changes in his mindset that he himself clearly noticed. Later, he was given a "bedstead" with an "orthopedic mattress". He would also be given medical support for his back pain. Naturally, time was passing. 7 (seven) years had passed since the former Leader came to the planet and started to enter the library. During this time, he had read many books and prepared presentations for them, and some of them were subjected to exams. Up until that point, the average score he received from all the exams and presentations was 62 (sixty-two). However, there were also times when he had rewritten many books (for the same book) and given presentations and taken exams in the same way 3-4 times. In the past 7 (seven) years, there had been significant improvements in his living conditions compared to the beginning. This situation at that moment was as follows: "He was given a 1+1 apartment on the 4th (fourth) floor of a twelve-story apartment building." Even though the living conditions in this apartment were not 100% like the other people living on the planet, it was quite good compared to before. The former Leader was happy with his situation. However, despite receiving all sorts of medical support, his back pain still persisted. Sometimes it hurt so much that he even had difficulty walking and standing. Because of this, the following conversation took place between him and the robot that had given him the task of reading a book in the library;

Robot: "Yes, please state your request."

Former Leader: "As I mentioned before, my back pain is still ongoing. For this reason, I would like to make the following request:"

Robot: "Yes, what is it?"

Former Leader: "I also want to read books about medicine. Maybe this way I can both relieve my own back pain and be useful and helpful to those who have similar troubles and problems."

Robot: "Okay, okay. There are already many books about medicine on your book list. With this request, you are moving the order of those books a little bit earlier. In other words, reading medical books, preparing presentations and taking exams were already included."

Former Leader: "Okay then."

Robot: "Now, we will take a 2 (two) week break before moving on to the next book. However, in the meantime, you can do whatever you want. You will travel around our planet accompanied by 2 (two) robots and 2 (two) humans. In other words, you will learn about our planet from its nature to its history. This will help you better understand the books you will read later."

Former Leader: "Okay."

Meanwhile, the people living on the Former Leader's planet were very pleased with the living robots that took over the administration in the last 8 (eight) years. However, the robots wanted to continue their rule for another 3 (three) years and then leave the administration completely to the planet in question. They would only assign a limited number of robots as "supervisors/controllers/controllers." They explained this situation to the public in detail through the press on the planet. The people were pleased with the administration of the living robots. Some of them even said: "After you leave the administration, we are afraid that someone like our former Leader will come again and we will return to those bad days. Therefore, at least we and you should have a 50%, 50% say in the administration. In other words, half and half. How can that be done?" The spokesman of the robots on the planet responded to these words by saying, "We understand that this is the wish of a very large percentage of the people living on your planet. Since you insist and want it to be like this, then of course it can be, why not," he said. The necessary preparations were made on the planet and then an election was held. The "Republican Regime" was chosen as the form of government. In the election, 750 (seven hundred and fifty) Deputies entered the National Assembly of the Planet, and the same number of living robots would serve as "Deputies upon Request". Thus, the distribution/number of Deputies in the National Assembly of the Planet was half and half. The new Assembly began its duties in a short time. The Prime Minister and the President were from the planets. There were living robots among their deputies. Thus, the process was proceeding normally and naturally. In one of the parallel universes that bordered the parallel universes where these planets were located, there was the following development: A competition was held in a galaxy in this parallel universe in which 2 (two) planets participated. The content and subject of the competition was as follows: Whichever planet could form/create the largest volume and most massive celestial body in the region whose coordinates were determined within 1 (one) week (according to Earth time), that planet would be declared the winner of the competition and all the people living on that planet would be given one of the most advanced spaceships of that parallel universe – one for each person. The competition had started 2 (two) hours earlier. The competitors were trying to take any celestial body they could, regardless of their science and technology, from both their own universes and other parallel universes, to the coordinates in question and transform them all into a single celestial body here. The celestial bodies they had taken and received included stars, planets, large and small meteors, etc. Since there were not many celestial bodies in the parallel universe they were in, they were taking/ pulling from the nearest parallel universe. The process was going on like this. It was wanted to include the planet where those people and living robots were located in this process. It was noticed that the coordinate information of this planet had changed. The reason for this was noticed and found immediately. There was a situation like the 2 (two) opposite poles of a magnet attracting each other and the planet was advancing at an increasing speed in the universe it was in. Nobody knew where and why they were going. However, a reaction was given to this situation that developed immediately, suddenly and unexpectedly. The planet adjusted its direction and direction of travel by itself and entered the coordinate region where the planets in a star system where celestial bodies were very dense and it was very difficult to be pulled or exited when entered. After entering here, the effect of the force that was pulling it was not felt on the planet at all. In other words, the force that wanted to pull the planet was far from being at a sufficient level in this new region of the planet and therefore the attraction of the planet was no longer possible. However, many other planets were not so lucky and they were rapidly moving towards that parallel universe under the influence of the gravitational/ reception force. In this way, 15 (fifteen) celestial bodies were about to enter the parallel universe where the competition was held by being drawn. There was also gravitational and reception from other parallel universes. Finally, the duration of the competition, which was 1 (one) week, was completed and one of the planets naturally won the competition. The volume and mass of the celestial body that enabled them to win was close to the mass and volume of the Milky Way Galaxy that we are in now (20XX's). In other words, they had created such a large celestial body. As a reward, each person living on that planet (those who were older than a certain number) was given one of the most advanced spaceships of that parallel universe. Each person who owned the ships set out on a journey of discovery with these ships, which could travel very long distances using little fuel, in order to see and find different things. First, they would travel in their own parallel universe, which is where they were, then they would go to the inter-universe region and from there they would go to the universe where humans and living robots were located. This was the plan they made for the trip. Soon, they began their journey of discovery with their spaceships with billions of planets.

Some of them did not go to their own universes but went directly to other universes. All of them naturally had electronic communication with each other and their planets. Those who went out of their own universes entered the universe where humans and living robots were located. In the meantime, the planet of the living robots that had previously changed places was brought back to its original place (its old place). There were those who opposed this, but what they said was not valid. Because the coordinate region they went to (in order to be freed from gravity) was very close to a huge star and therefore very hot. Therefore, they could stay there for a maximum of 5 (five) months and then they took the planet back to its old place. In the meantime, the spacecraft that went on an exploration journey reached the atmosphere of the planet where the living robots were located. The spaceships could travel much faster than all the spaceships in the known parallel universes. For example; 10 light years for a normal spaceship could be like 0.5 light days for these ships. In other words, they could travel a distance that took 10 (ten) years at the speed of light in 0.5

days by traveling much faster than the speed of light. There was such a big difference in terms of speed. Approximately 10,000 of those spaceships came into the atmosphere of this planet for exploration purposes. The robots that saw this unexpected number of spaceships on their receivers immediately went into the highest level of "alert" status. Because it was the first time they saw so many spaceships at the same time and they did not know the purpose of their arrival. Contact was established with the ships immediately and they were asked why they had come. The answer they received was "they were traveling, they were on an exploration trip." The people who came to the planet told the living robots that they had previously moved the planet to pull it. Thereupon, the people on this planet and the living robots wanted to make these uninvited visitors do something. First of all, they were all told "welcome." After they landed at the landing base of the planet with their ships, they went out. The living robots that appeared next to each of them the moment they came out said to each of them "we will punish you, since you wanted to pull the planet we are in into your own universe, we are giving you, each of you, the task of sweeping and cleaning a separate street of ours." The necessary materials were quickly prepared by the robots and the uninvited people began to wash and sweep the streets, one street at a time. The robots stood over them, watching them and giving them instructions. Sometimes they would say, "You skipped this place, it wasn't clean, wash it again, sweep it again, etc." Finally, many streets were spotlessly clean and sparkling. Afterwards, the cleaners got into their spacecraft and left the planet. In the meantime, the people in the spacecraft, who couldn't stomach having their cleaning done, talked among themselves and prepared a plan. According to this plan, a living robot from the planet would be tricked into getting into a spacecraft and then kidnapped as a punishment for the planets. For this purpose, someone from one of the spaceships said, "There's a problem with our ship, it's not moving. Can you send someone to check?"

Robot: "Okay, okay. It'll be there in a minute."

The robot assigned for this purpose was brought to the spaceship in question with a spacecraft and docked. Then the robot moved to the spaceship. As soon as it passed, it was thrown at him. His hands and feet were immediately tied and then he was knocked out. Those who wanted to kidnap the robot did not know the many and different features of the robot or what it could do. Although the robot's hands and feet were tied and it was unconscious, it took the shape and structure of a very thin and very long pulley rope and tried to escape from where it was by crawling like a reptile. However, 5 (five) people on the ship noticed this situation. They threw a transparent cover on it. Then they folded and shrank the cover and poured it into a transparent but metal jar. Thus, the living robot, which had become as thin as the pulley rope, was imprisoned in the jar. The people on the ship were saying, "Okay, this will stay in this jar until it dies. I hope it was a good lesson. There is no point and meaning in keeping it on the ship, let's throw it into space from one of the windows of the ship." They did as they were told and threw the transparent metal jar out of the spaceship. At this moment, the living robot inside the jar changed shape again and this time it started to grow inside the jar, that is, to increase its volume. Then, as it grew even bigger, the metal wall of the jar cracked and broke into metal pieces. Then, the living robot was freed from the jar it was in. However, there was no oxygen where it was. If help did not come, it could stay there for a maximum of 45 (forty-five) minutes. This period had started and exactly 25 (twenty-five) minutes had passed when a spaceship that had come from their planet to take it away approached it. The robot, which immediately entered the spaceship, started breathing deeply here. Thus, the living robot was freed without dying like a living creature we know. There was a planet in another star system not too far from the star system where the living robots were located. This planet was constantly under attack. For this reason, the ruler of the planet was tired of both the attacks and dealing with the people's problems. He and a group close to him prepared a plan. This plan was as follows: "A very large shelter had been built on the planet before, to be used in times of attack or war. A new attack on the planet would be expected, and when there was an attack, all of the planet's population, which was not very large anyway, would be asked to enter the shelters, and after entering the shelters, the roof would be closed with equipment that could not be opened from the inside, and the people would stay there, in the shelter. They would probably die in the shelter because they would not be able to get out. After the roof of the shelter was closed, the administrator and the team close to him would leave the planet with spaceships and go to another planet. A new and powerful attack was expected. The planet was already under constant attack, but these were not constant and were of varying strengths. Such a moment came and a very effective attack began. According to the plan, the people were asked to enter the shelters. This process took approximately 7 (seven) hours. When the shelter was being built, it was designed in such a way that it could be stayed in for a maximum of 1 (one) month. After everyone entered the shelter, the roof was closed and the people inside were practically imprisoned. Then the administrator of the planet and his relatives left the planet with spaceships. The planet now resembled a celestial body, with no living species other than animals and plants. Before leaving, the administrator and his team had removed the planet from its orbit in the star system it was in. For this reason, the planet was moving through the universe, passing randomly to random coordinates and bumping into celestial bodies as it passed. Thus and in this way, it continued on its way for about 2 (two) weeks. After the said period, it entered the star system of the planet in which there were living robots. Right after entering here, the random progress of the planet was corrected and it started to advance in a regular line. In the meantime, the people still in the shelter realized that the shelter was closed from above. No matter how hard they tried to open this hatch, they could not. In order to make it noticeable that they were there, they wanted to use the ventilation lines that connected the shelter to the outside. They lit a fire inside the shelter, structured the smoke of the fire in different colors in order and ensured that it spread through the ventilation lines towards the upper parts of the planet, that is, towards its atmosphere. In this way and in this way, they thought that someone would notice the situation and come to save them. In this way, 5 (five) more days passed. Since the fire was constantly lit inside the shelter, the amount of oxygen inside continued to decrease rapidly. Another 3 (three) days passed when this planet became tangent to the planet in which there were living robots. At that moment, colorful and continuously spreading smoke was noticed. Since it aroused suspicion, a spaceship was sent to this planet to conduct research and examination. The ship landed on the nearby planet within seconds. After landing, 30 (thirty) people on the ship immediately walked to the area where the colored smoke was coming from. When they approached that line, they heard not only colored smoke but also sounds coming from there. Thus, they understood that there were many living beings where the smoke was coming from. The hatch on the top of the shelter was open and clearly visible. There was a latch next to the hatch. It was made so that the hatch could be opened easily from the outside. Immediately, one of the robots lifted the latch upwards and at that moment, the large hatch slowly began to open. Then, everyone inside the shelter climbed up the stairs used for the people to go up and everyone inside came to the surface. Since they were very exhausted inside and most of them were sick, it took almost 12 (twelve) hours for all the people to get out. When the people's spokesperson went out, they came face to face with the robots that saved them. They talked to each other and learned about each other.

Robot: "If you want, let us include your planet in our star system. Instead of traveling randomly, you will be one of the planets of this star system."

Public spokesman: "Of course."

Later, after many procedures, the planet in question was included in the planets of that star system. The majority of the people of the planet were engaged in agriculture. Therefore, as a way of thanking them, they presented the planet where the living robots were located with many melons and watermelons with their own characteristics. Their geometric shape almost resembled a pentagon. In other words, they were not elliptical, spherical or angular, but had a pentagon-like shape. The gas that was previously stored on the planet Venus and then started the formation of life there had the following development: The gas here had more or less leaked and progressed through the void of space and reached Mars. Some of it had also headed towards Mercury. However, since Mercury was very close to the Sun, the gas that reached here had lost its characteristics and the structure that would initiate life species. The part that reached Mars had already started the formation of different types of living beings here after all this time, and even living beings in this sense had already started to multiply. The people on Mars (who came from Earth) finally came face to face with the living beings that were formed by the effect of the gas before too much time had passed. So much so that, within these living beings, there were living beings of very different types and structures, such as a mixture of giraffe and lion, a mixture of crocodile and elephant, etc. Both in the air, on land, in water and under water, billions of different different species. Some of these animals were occasionally attacking and even killing people walking around Mars or going to work. The Martians wanted to find a solution to this. Thousands of years ago, distinguished scientists from Earth first went to Mars, but later on, many terrorists, murderers, thieves, etc. who were involved in at least one crime on Earth and committed a crime, were wanted but could not be caught, escaped to Mars en masse and in groups. The state authorities on Earth at that time had not implemented an operation to catch these criminals and had not implemented such a thought as to go to Mars and catch them there. Thus, on Mars, people who were both very advanced and very backward in terms of thought – in terms of positivity – were living on the same planet. And there were also wild animals that appeared unexpectedly. The people on Mars had Earth to ask for help from. However, Earth had long since disintegrated and disappeared and scattered into space. They were also aware of another planet where living robots and people lived, but they refused to ask for help from them. They thought, "It would be best if we took care of it ourselves" and wanted to implement this. They thought of taking action to do something in this way. However, they had very limited means – in terms of tools, equipment, materials, equipment, etc. They had weapons, but they did not have a certain or regular army. They only had different security units consisting of a limited number of personnel. It did not seem possible that these could combat the wild animals in question. The Martians first thought of placing the animals they considered most dangerous – if they could catch them – in a large zoo they had built. To do this, they stunned and caught tens of thousands of animals that they had previously decided to catch, and took them one by one to the zoo. There were also many birds and bird-like animals of different sizes that had originated on Mars, but they had not exhibited any harmful behavior up until that point. Therefore, the Martians did nothing against these flying creatures. On the other hand, in a galaxy very close to the star system where living robots and humans were located, the process of a giant star becoming a black hole was progressing very rapidly. The process of turning into a black hole was progressing so rapidly and towards such a large region of the universe that the black hole overflowed into the inter-universe region, in other words, it extended all the way to there. Due to this development, the planets neighboring that galaxy entered the black hole and exited the inter-universe region. In other words, many planets and many celestial bodies suddenly found themselves in the inter-universe region. This included the planet where the living robots were located. There were many other planets that contained life. (It passed to the inter-universe region.) This region was irregular, that is, it showed different physical structures at different time intervals. For example; 20 (twenty) years liquid, 5 (five) years solid, 15 (fifteen) years gas, 2 (two) years nothing, etc. structures. 2 (two) months before the planets came there from inside the black hole, they stopped behaving like liquid and started to show normal properties. However, as with every feature, no one knew how long this feature would last except God/Allah. It was like a computer program generating random numbers, it was showing random features or behaviors at random times. It was not possible for them to know how long this nothing feature would last without living for that moment. However, they were lucky because they had not come to the interuniverse region when there was neither liquid, gas nor solid feature. They were aware and conscious that they had to leave this region, which was not suitable for them to live, as soon as possible. They were thinking about what they could do for this purpose. The devices on their planets stopped working after they passed to this region, including power outages. Each planet stood as fixed as a nail hammered into the wall in this region. There were exactly 150 (one hundred and fifty) planets and 20 (twenty) stars and countless other celestial bodies in this situation, that is, passed to the interuniverse region. The stars that passed here were gradually losing their star feature. In other words, each star was gradually fading (?) and progressing towards becoming a planet. Since there was no electric current, they had no possibility of emitting devices that could work with the effect of stars and escaping from the inter-universe region with their repulsion power. However, there was something and they realized it. This was this: Even though it was an inter-universe region, the gravitational forces that existed due to the masses of the celestial bodies here were effective and active. Here, living robots and humans and also other planets with other living beings began to use these gravitational forces to find ways to get out of the inter-universe region and escape. The devices that worked with magnetic effects continued to work. Using these, they also included other planets that did not have this technology in this rescue and started to move from the inter-universe region, albeit very slowly. All of the planets were moving, in other words, the process of escaping had begun, but they were moving very slowly. There was no concrete data on which coordinates the planets in this situation were in the inter-universe region and how far they were from the nearest parallel universe. Therefore, all of the planets were moving slowly, not knowing when they would get out of there but considering it certain that they would. As they moved away from the stars there, the gravitational force gradually lost its effect and for a while the planets started to move even slower. In an electronic sense, the planets had no communication with each other or with parallel universes. In a different sense, there was a magnetic field spread in the inter-universe region. However, this situation was very, very foreign to the scientists on these planets. While they did not yet understand this magnetic field, in order to communicate using it, a process similar to the time elapsed from the beginning of electronics to that moment was required. Therefore, it was not possible to use it or produce a solution to provide communication in the short term. 5 (five) years had passed since the beginning of their entry into the inter-universe region and the planets that were in motion had still not left this region. In the meantime, due to unsuitable living conditions, all the people and natural living beings (animals, plants, etc.) on all the planets lost their lives. Only one planet remained alive. They were living robots. In other words, apart from these, not a single living being remained on any of the planets that entered the inter-universe region. Thus, the human race and other natural living beings generation or process was completed on these planets. However, the natural generation of living beings continued naturally on the planets in parallel universes. Despite the long period, there were no problems with the living robots other than some very minor negativities. Another 3 (three) years had passed when suddenly the electric current was obtained again. However, at this moment, the effect of the stars whose gravitational force they used was reset on the planets. After the electric current was obtained, the equipment working with this current became operational and all the planets started to travel at very high speeds to leave the inter-universe region as soon as possible. So much so that the planets were moving at speeds close to the speed of light. They left the inter-universe region approximately 4.5 (four and a half) years later. However, the stars that remained there were still in their places, that is, they continued to stay in the inter-universe region. The planet hosting the living robots entered a parallel universe, but the information they obtained from there was that this universe was not the old universe they were in. Since the devices working with electric current were now active, their distance to that old parallel universe they were in was calculated as trillions of light years. Therefore, they decided that it was no longer possible to go to their old universe. They started looking for a galaxy suitable for their planet and their own structure in the new universe they had entered. While they were advancing there, they saw a star system with 2 (two) stars. The planets here had very interesting orbits. Each one first completed its rotation around a star and then moved into orbit around the second star and completed its rotation around it. They saw this situation in their receivers and decided to enter this 2 (two) star star system. When they entered, the planet was shaking, in other words, it was shaking. After advancing for a certain period of time, its shaking stopped and it entered a certain orbit. According to the calculations, it would complete its rotation around the first star in that star system in 3 (three) years (Earth years) and its rotation around the second star in 7 (seven) years (Earth years). In addition, it would complete the transition between each 2 (two) orbits, that is, the transition from one orbit to the other, in 7 (seven) months (Earth time). The orbit from the first orbit to the second orbit and from the second orbit to the first orbit was the same and the transition period was 7 (seven) months. (Earth time). In this case, again according to the calculations made, while they were around the first star, they would experience each of the 4 (four) seasons much longer than on Earth. In other words, the duration of the seasons would be long. While they were around the second star, they would experience the 4 (four) seasons again 3 (three) times this time. In other words, each season would be experienced 3 (three) times. Therefore, the transition between seasons would be faster both compared to Earth and the orbit around the first star. The living robots that started a new life in these new universes and new galaxies missed their old universes and star systems. They were satisfying their longing and yearning in another sense by watching the audio videos shot from those days. They had not come across any living species or even a single living being in their new universe. They were asking themselves, "I wonder if there is any other living and natural life in this parallel universe besides us?" They did not wait long to get the answer to their question. Suddenly, a spacecraft that covered that shape spherically arrived in the atmosphere of their planet and a communication link was established with them.

Robot: "Who are you, what do you want from us? Why did you include our planet in a global frame?"

: "The real question is to you. Did you take it from us before coming to this universe? Did you come and settle on your father's farm without asking anyone?"

: "We did not come here willingly and knowingly," he said and explained what had happened.

: "Okay, I believe you, we think you are not lying. You cannot stay in our universe permanently. This universe belongs to us, the natives of this place. We will send you back to the universe you came from, whether you want it or not."

: "We wanted to do this too, but we came here because we could not return to our old universe. If you can send us to our real universe, we would like to give you a large number of our most advanced computers as a gift. So how will you send us to our old, real universe, do you have that technology?"

: "If we did not have it, then we would not be talking about – we can send it – anyway. Now give us a large number of those computers you mentioned earlier because the journey to your universe, which will take 3 (three) years, will begin soon."

: "Okay, we don't think you're lying either. The computers are already ready, you can come and get them."

: "Okay, right after you get all the computers, your journey to your real universe will begin."

: "Okay, we have an agreement."

After the advanced computers in question were delivered, the planet's long journey to its own real universe began. Finally, the time was up and they reached their own star system in their own universe. However, there was neither a planet nor a star where they arrived. The coordinates of the old star system were empty. The giant black hole that had previously caused them to go to the inter-universe region had also swallowed the planet and star here. Thereupon, the search for a suitable star system began. They discovered a star system that they had never known or been aware of until that moment. Not only was the star here radiating warmth, but also both heat and cold air, snow and rain. They also saw such a star for the first time there. However, according to their predictions, they said, "A normal or natural star would not behave like this, this star must have been exposed to unknown influences." They positioned their planet in a suitable orbit among the 12 (twelve) planets here. There were already 4 (four) seasons on the planet. In addition to this, in addition to the snow and rain that had fallen and would fall on the planet, there was also the snow and rain that would come from the star. However, since they could change the planet's orbit and movement as they wished, there would be no question of any negativity in the future. The star's color changed between blue and red depending on the situation. In other words, it also changed color according to the effect it had on its surroundings. When it radiated heat, it was red, when it radiated coldness, it was blue, when it rained, it was light blue, and when it snowed, it was a lighter blue. The living robots on the planet felt that something was missing since there were no humans or natural life there anymore. For this reason, they were talking among themselves, "We don't know where we will find it, but let's find and bring humans and animals to our planet." For this reason, they set out on a journey to find natural life with numerous spaceships and numerous living robots. 5 (five) weeks had passed since the journey began, but they couldn't find anything they wanted. However, they were all extremely determined and confident. They thought that "there must be at least one planet where natural life as we know it lives" in the universe whose boundaries they didn't know. Another 5 (five) weeks had passed, but the result was zero to zero. At the end of this period, they saw very, very small spiders on a small meteorite, not a planet. How could they live on that meteorite? They didn't have any information or guesses about this. The meteorite looked more like a cylinder with a diameter of 200 meters and a length of 750 meters. The surface of this structure was literally crawling with spiders. The spacecraft was lowered onto this meteorite. They saw that the spiders were gnawing and eating the meteorite they were on. In other words, their food source was the meteorite they were on. Many pieces of the meteorite were taken and examined in the laboratories on the spaceships to check what was in it. During the checks, they realized that the meteorite was like a giant dried and hardened bread. It was the first time they had come across a meteorite with such a structure and content. They said, "Since spiders eat it, then it must be healthy." They caught as many spiders as they could and put them in the previously prepared section on the spaceship. They decided not to continue their journey for the time being and said, "These spiders will be enough for us at first, we will spend time with them and then when we get tired and bored, we will continue our journey to look for other creatures." Hundreds of thousands of spiders were brought to the planet by spacecraft. However, none of them were released and released into the nature of the planet. They had built a place similar to a zoo and placed the spiders there. In this way, they would all continue their lives under control. After 5 (five) years, the journey to find natural life started again with many spaceships. During the journey, they encountered a very different meteorite. Its color was constantly changing, including the basic colors of red, green and blue and their shades. A spaceship wanted to land on this meteorite and see its structure from above. However, when the spaceship landed on the meteorite, the meteorite swallowed the spaceship, that is, sucked it in. The living robots on the spaceship, who watched these events from a certain distance from the meteorite, were shocked. Because it was the first time they had encountered something like this until that moment. - There are some plants that close their leaves when any fly, insect, etc. lands on them and sucks it in - This meteorite's behavior was exactly like this. Other spaceships showered it with laser beams in order to find the lost ship or take it out of the meteorite. Very powerful laser beams were applied to the meteorite for very long periods. As a result of this effect, the meteorite broke into pieces as small in volume and mass as grains of sand. However, the spaceship it swallowed was nowhere to be found. Where did it go and how did it disappear? They could not find the answer to this question. The living robots in the other spaceships waited sadly for a long time next to the meteorite that had turned into sand. Then they decided to continue their journey where they left off. The journey continued for weeks but they did not encounter any living beings. They did not encounter any living beings on any planet, but they noticed a spaceship whose direction was parallel to the direction of movement of their own spaceship and the same direction as their own direction (arrival). The alien spaceship was getting closer to the ships. So much so that when the distance between them was 250 (two hundred and fifty) meters, the spaceships and the alien spaceship stopped. The conversations they had with the electronic communication devices on their ships were as follows;

Robot: "This is the first time we have encountered a spaceship other than ours, who are you, where are you coming from and where are you going?"

: "We are asking you the same question as before?"

Robot: "Will you answer or shall we answer?"

: "Okay, let us answer. We are chasing a terrorist who escaped. He bombed all the factories and production facilities on our planet. Most of them are either out of service or no longer usable, that is, they can no longer produce. Just as we were about to catch the terrorist, he boarded one of the fastest spaceships on our planet and escaped at full speed. We have been following his trail for exactly 5 (five) years. We find his trail from time to time, but he finds a way to lose his trail again. That is why we are on the road. What about you?"

Robot: "We are not natural creatures like you, we are living robots. We set out on a long journey in our universe to find natural creatures. The last time we saw spiders, now we saw you. We have not seen or encountered any other natural creatures."

: "Well, we will give you the coordinates of the galaxy and planet where our planet is located. If you go there, there are billions of natural creatures like us. You can communicate with them. Now we will continue to search for the trace of that escaped terrorist. We have already lost too much time by standing here. Goodbye for now."

Robot: "Okay, I hope you catch him."

After receiving the coordinate information of the galaxy and planet with natural life from the spaceship they encountered, they set off to go to this new region with one less ship. The journey was relatively short and was completed in 7 (seven) weeks. This place was literally teeming with natural life. There were at least 20 (twenty) billion natural life on each of the 4 (four) planets in this star system, more than 150 (one hundred and fifty) billion natural life. There had been a terrorist attack on one of these that they knew about beforehand. They had received this information from the spaceship they encountered. They learned which planet this was and landed there with their spaceship. After explaining the situation to the authorities who welcomed them, they stated that they wanted to see the factories and production facilities that were inoperable and badly damaged after the attack.

: "Why do you want to see it, does it have a special meaning?"

Robot: "No, not because it has a special meaning. We want to see if we can get them back to working order."

: "Okay, understood. Then let's go and see. But there are tens of thousands of them. It would take a long time to go and look at all of them. That's why we will go and see the ones that are important to us."

Robot: "Okay, but how do you determine the importance or priority?"

: "According to the importance and value of what the living beings on our planet produce for their continued existence. We determine the priority accordingly."

Robot: "Okay, understood."

The robots and the natural creatures walked together and sometimes got on land vehicles and visited the factories and production facilities in question one by one. While they were visiting, the living robots took plenty of notes, took both photos and video footage and recorded what the people there said. In this way, they were only able to visit important facilities in 5 (five) months. Then they prepared a long-term report and gave it to the planet's representative. The report stated that "while it was stated that production in some production facilities could start in a very short time (such as 1-2 weeks), in some there could be a repair, maintenance and commissioning period that could last up to 5 (five) years." They explained that the number of those that could be operational again immediately, in a short time, could be around 45 (forty-five). Among these were facilities that produce in a very wide range from the food sector to the health sector and the electronics sector. Production started in 1-2 weeks with the work carried out by the living robots with the equipment they brought with them, as they mentioned before. For the repair and maintenance of the others, the living robots had to bring a large number of tools, equipment and materials from their planets. This meant that additional time was needed to get these from there. However, they did not go to their planets, electronic communication was established and information about the situation was given. Then, whatever was needed was set off to be delivered by 5 (five) spaceships. During this expected period, the living robots learned about technologies that were not on their own planet but were on this planet. Among these technologies, the first was "spacecraft engine production". The next was "facilities that produce in the field of health". Especially those related to the field of "health" were the first things they saw for living robots. Although they were "living robots", they were "robots" after all. The medicines they produced attracted their attention. The difference and change they saw in the production of "engines" was that there was "no living being – namely humans – contribution". The requirements were determined autonomously with "artificial intelligence" completely and the design was made accordingly and then production was started. In addition, since the population of this planet was expressed in billions, there were many spacecraft. Therefore, a large number of engines had to be produced very quickly. So much so that, as in the parallel programming logic, production was done in a similar way. Engines were produced, tested and then made usable in a time that could be expressed in minutes. These engines did not use any matter or various physical states of these matters as fuel. Their fuel was the gravitational forces of celestial bodies themselves. They could use the gravitational forces of all celestial bodies they passed by or at a certain distance as fuel for their spacecraft. This meant that "there is as much fuel as there are celestial bodies in the universe". In other words, if these spaceships did not go to the inter-universe region, there would never be a fuel problem. Finally, 5 (five) spaceships brought everything needed for the repair, maintenance and maintenance of the production facilities to the planet in question. These were grouped in advance according to their features and where they would be used and arranged in a way to form a unity. Due to the software and hardware structure features they contained, these incoming hardware had the knowledge/information of where they would go and what they would do, how long they would work, how long they would wait, etc. everything that came to mind on this subject. Although it was previously estimated that it would be completed in 5 (five) years, the period did not take that long and in 3.5 (three and a half) years most of the factories and production facilities were restored to their former structure. In the meantime, when this period was completed, they were able to catch the terrorist they were after. The terrorist was brought to the planet in a transparent capsule. Then he was put in prison for the rest of his life. When the terrorist was asked, "Why did you do such a thing?" his answer was: "The manager of one of the factories did not give me the way while I was walking on the road and stopped me by getting in front of me. I bombed them all to get revenge on him." For such an ordinary and simple reason, countless factories and production facilities were damaged to the point of not being able to operate. There were still facilities that could not be restored to operating condition. The reason for this was that very effective bombs were dropped on these places and because of this, not only the equipment inside but also the buildings were blown up to their foundations. In other words, the place where the facilities were located was shattered into pieces. For this, the equipment brought in was first recycling the piles found here, producing new construction materials from the demolished ones and in other words rebuilding the facility from scratch. This process took a little longer than the others. Since the population on the planet was very high, there was also a water problem here. Moreover, there was no ocean on this planet. The number of seas was expressed with the value of a single-digit number. There were many lakes and running water of various sizes. They had a very hard time obtaining fresh water sources, namely drinking water. They were not bringing this water from their own planet, but from another planet. The other 3 (three) planets in this star system were bringing drinking water from here. Thus, production in the bombed facilities had started and continued, even if it was not as much as before. In the meantime, a distress signal call came from a planet to the planet where living robots were located. The words of the person asking for help were heard in the receiver as follows: "Those who are in the administration of our planet, except for their own lineage, are executed when they reach the age of 35 (thirty-five), regardless of who they are, whether male or female. The reason for this is that they see themselves as God/Allah and say - it is enough that you have lived this long, we cannot allow you to live longer than us. If you do not come here and save us from these, the executions will continue. This ruling dynasty has been acting like this for 450 (four hundred and fifty) years. I reached the strongest transmitter on the planet by chance, taking advantage of an officer being distracted, and I am sending this message. I wonder if it reached anyone? If it has reached, I will specify the coordinates of our planet in a moment, have it come here and save us from this situation very urgently." This audio signal (there was no image) was recorded on the planet where the living robots were located and the coordinate information of the planet was obtained and preparations were made to start the rescue operation. The information received did not include information such as who and what kind of war or rescue operation would be fought, the number of soldiers and weapons the administration had, what kind of weapons they had, etc. Therefore, the living robots made an estimated calculation and immediately set off with spaceships to conduct a rescue operation with a certain number of living robots and weapons. The planet at the mentioned coordinates was reached in 1.5 (one and a half) years. When they went, they saw that the technology on the planet was much backward compared to theirs. The ruling dynasty had millions of soldiers, but the weapons these soldiers had were very primitive. They had weapons almost like slingshots. Therefore, the living robots almost without any difficulty neutralized all of the soldiers of the dynasty in question and forcibly put the dynasty (each one) into the spacecraft. The total number of members of the ruling family was 2,500 (two thousand five hundred). They were all placed in 3 (three) different spacecraft and were to be taken to a planet they did not know but was suitable for life and left there. The people who were saved thought the living robots were natural creatures and said to them, "You govern us from now on." However, the living robots explained that they and natural creatures were not the same and the difference was there. They also said, "You govern yourselves better, not us." Thereupon, the insistent people gave up their insistence. A person from the people asked a living robot

: "What did you do to those in power?

: "Don't worry, we did not execute you like they did to you. We took them to a planet they did not know before but where they will continue their lives. Let them see what they are like there."

: "Okay, we are very happy. So, won't you ask us for anything for saving us?"

: "There is nothing we particularly want or want you to do. However, no matter how many people from your planet want to come to our planet and stay with us, they can come with us."

: "Okay, I will tell everyone. But the result will be clear in at least 5-6 weeks. Can you wait that long?"

: "We will wait, no problem."

: "Okay then."

The time in question has been completed.

: "I took down the names and surnames of those who want to come. It took pages and exactly 7,500,000 people want to come to your planet."

Robot: "This number is too many. We cannot host that many people and our spacecraft would have to make countless trips to take them all."

: "So how many people do you plan to take with you?"

: "There will be a maximum of 250,000 people."

: "So how will you determine who these people will be?"

: "We will record the names and surnames on the paper you brought as data to the computer by assigning them a number. Then a program will run and randomly determine 250,000 of them. They will come to our planet with us."

: "So there will be people who will go from our planet to your planet. Do any of you want to stay here with us?"

: "That is not possible because we need to be together due to the technical features we have."

: "Okay, okay then."

The information was loaded into the computer's memory and the program was run. Then, the name and surname information of 250,000 people became clear. Since no more than one person had the same name and surname, that is, both, there was no conflict of names and surnames.

: "What do we have on your planet that you don't? You've seen around, if there is such a thing/thing, let's take it with you."

Robot: "Yes, there is something that we don't have on our planet. That is some fruits that are on your planet. We want to take some of these fresh, their seedlings and seeds, is that okay?"

: "No way, of course it is."

Then the preparations were made and the journey to the planet where the living robots were located began. Thus, a new life began for 250,000 natural living beings. Whether there is life in the universes and the planets in the galaxies within them, there are definitely problems. If there is life, the problem is noticed. (On a living basis.) If there is no life, the problem maintains its naturalness within the nature of that planet. (In the sense of being solved, eliminated or not being solved and eliminated.) In this way, while the process was flowing or progressing, another call for help reached the planet where the living robots were located. This time, the call was not only auditory but also visual. In other words, there was a visual and audible call for help. There was a recorded video in the incoming image. It showed 7 (seven) planets and small celestial bodies that were attached to the land of each planet and had star characteristics. It was understood from these images that each of the 7 (seven) planets had its own star on it (attached to its land). There was a vital problem that needed to be intervened immediately for 3 (three) of these planets. The stars here had started to send out extremely high heat to the planet they were in. Because of this, the number of deaths among the people living on the 3 (three) planets in question continued to increase every day. The situation had to be intervened immediately; either those tiny stars had to be removed from the surface (land) of the planets or the heat they were emitting had to be prevented. They had not received any response up until that moment. However, since the planet with living robots periodically scans its receivers for different frequency regions and this scan time coincides with the first planet's broadcast time, the signals sent were received. Thus, the first planet achieved its wish to transmit the signal. The response was given immediately and it was said, "Okay, give me the coordinate information of your planet, let's set off immediately." After the necessary information was received, many spaceships with many robots, weapons and other equipment set off from the planet where the robots were located. According to the coordinate information given by the first planet, they could reach this planet with the spaceships they were currently using in at least 4 (four) years. Although this period was insignificant for living robots, it was an important period for natural creatures. The journey continued and nearly 4 (four) years passed. The coordinates where the planets were located were reached. Just as they arrived, the spaceships landed on the third planet, the planet where the Çx element was extracted. Living robots and weapons were placed here and it was made active, in other words, usable. In line with the decision taken by the living robots, information boards were placed in many places on the third planet. Some of these boards contained the following information: "We came here from another galaxy. Each of the 2 (two) planets was given equal periods of time to extract the Çx element. This period is 2 (two) weeks. After the period is completed, the other planet will come and operate to extract the Çx for 2 (two) weeks. While one planet is performing the element extraction function, the other planet will definitely not interfere in any negative way. If they are found, they will find us in front of them, we will immediately intervene, resist and repel their attack. In short, this is the situation." In addition, this information was written in the air with laser beams in variable colors that could be read day and night. Despite all this warning and information, the second planet did not seem to be convinced. Again, they attacked while the first planet was trying to extract Çx. As stated in the warnings, there was an immediate intervention of living robots and all those who attacked on the second planet were caught and detained. In addition, as a punishment, the second planet was not given the opportunity to extract Çx for 8 (eight) months. When this period was over, the efforts to extract the element in question continued in order. Now, while the first planet was trying to extract the element, the second planet was not intervening. They had accepted the situation by force. In this way, their efforts regarding Çx continued without harming or obstructing each other. Living robots could not stay here for a long time to ensure peace, so they assigned electronic robots and then the work was connected to an automation. Then they left the planet where Çx was being extracted. Thus, while the flow continued, a new call for help reached the planet where the living robots were located. This call was the most different in terms of "subject of help" up to that point. The following was displayed on the receiver screen in audio and visual form: "The use of science and technology on our planet has increased and developed very rapidly. However, everything is out of control right now. Life on our planet has been turned upside down in every sense. Integrated circuits attached to living beings caused them to do things they had never done before. Wireless communication, electromagnetic waves, infrared, laser beam transmission, etc. everything became chaotic. Out of nowhere, spacecraft started to take off and bomb places, walking creatures started to fly, ships started to move on land, aircraft started to dive into water, etc. Every imaginable illogical and irregular situation started to occur on our planet. Computers started to produce absurd source codes in order to find another logical situation besides logical 0 (zero) and 1 (one). Frequency bands got mixed up, emergency calls were made on frequencies where radio broadcasts were made, amateur radio operators started talking on frequencies used by the army, etc. As I mentioned earlier, everything turned upside down. Living beings had already lost their natural living characteristics. There are dozens of integrated circuits in each living being. People started to attack each other for no reason and hug each other for no reason. It is as if a random chain of events is happening on our planet. If we cannot find a solution to this, our planet will lose everything, this structure that has already gotten out of control will now become an irreparable situation." Robot: "Okay, give me the coordinate information of your planet immediately. Let's come and do something for the solution."

After the information in question was received, a large number of vehicles in a well-equipped spaceship set off for the planet. The first thing they did when they arrived was to remove the integrated circuits attached to all living things. Then the electric current everywhere on the planet was cut off. Starting from scratch, the circuit diagrams and user manuals of all electronic/electromechanical devices were examined. The frequency band allocation and wireless communication were connected to a different standard than the existing standards. All equipment was cleaned and maintained. The difference between the laser beam frequency region used as a weapon and the laser beam frequency bands used in communication was increased even more. The activities and awareness of healthy nutrition, sports, education and training, fine arts, etc. were increased. New areas of interest that were unknown and not done until then were created. It was ensured that everyone definitely produced for nature, for society, for themselves, etc., and if they did, it was increased. Then, for exactly 1 (one) year, they were asked not to use any electrical appliances, devices, equipment, etc. unless it was necessary and vital. After these decisions were put into practice, the living robots decided to stay on this planet for a long time in order to see what was going on. After the 1 (one) year was over, everything that operated with electrical current was turned on. They were waiting with curiosity to see what kind of developments would happen. Two hours had passed, but none of the chaos that had occurred before had occurred again. The waiting period was increased even more. Another two weeks had passed, and again no negative events had occurred. Then, the living robots were asked, "What could be done for the patients whose integrated circuits had been removed from their bodies?" They said, "Let the newly produced integrated circuit be placed in the place where the old integrated circuit was placed." This procedure was carried out and these people who had entered the disease process regained their normal state. In the meantime, a discussion had started on the planet where the living robots were located: "Our receivers and transmitters are constantly open, everyone is asking us for help by telling their problems and troubles. Although we are not natural living beings, we are living robots. The universe we are in is not enough, and we are also dealing with the problems of other parallel universes. It is time to take a long break from this situation. Let's keep our receivers and transmitters closed for 3-4 years and cut off our connection with the outside world in every way and work to advance in science and technology on our own planet, that is, within ourselves. Therefore, this science and technology of ours will help those who ask us for help more than the current situation."

*Note: If we look at it with the logic of creation; now, every individual living in the Republic of Turkey has a Republic of Turkey Identity Number. The same situation exists in the universe we are in, in other parallel universes and in the inter-universe region. However, the equivalent of this has not yet been discovered. In other words, there is a feature that distinguishes the smallest structure of an existing volume (no matter what it contains or nothing at all) from the other smallest structures. What is this? There is no answer at the moment, but it is waiting to be found or discovered. This thing may not necessarily be a number, a word, etc., but it may also be connected to another concept that we do not know at the moment. If this discovery is made, it will initiate a much more effective and advanced structure than the invention of the transistor or the start of the space age...*

October 24, 2024 – Thursday, 19:30
Computer Engineer Metin Şahin

www.ingramcontent.com/pod-product-compliance
Lightning Source LLC
Chambersburg PA
CBHW021508210526
45463CB00002B/949

* 9 7 8 1 6 6 9 8 9 1 5 0 5 *